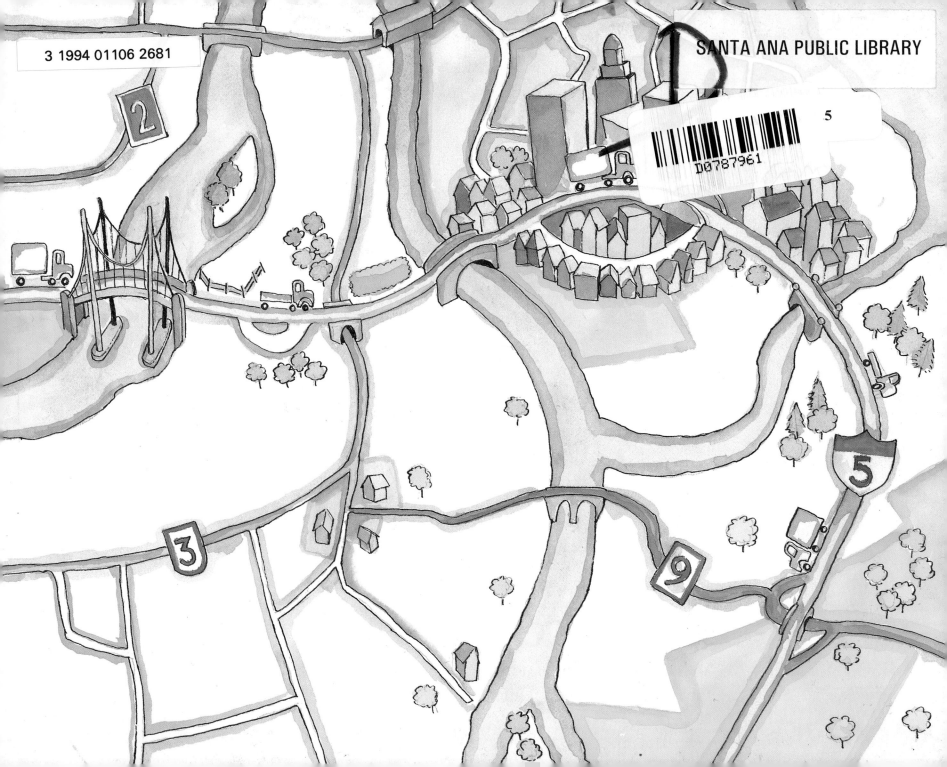

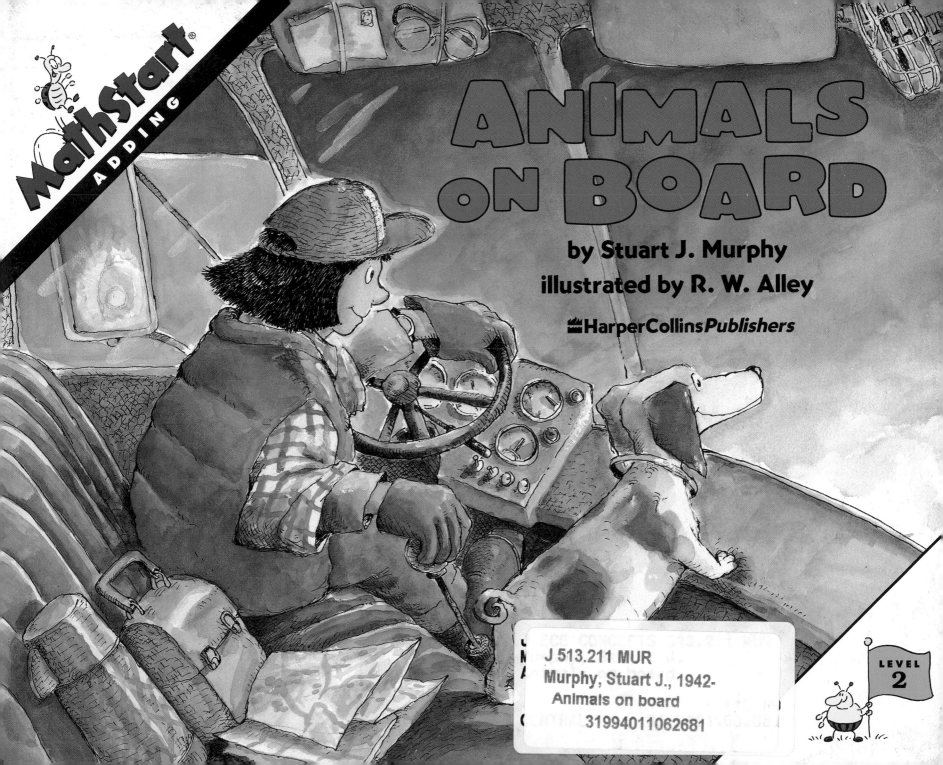

MathStart®
ADDING

ANIMALS
ON BOARD

by Stuart J. Murphy
illustrated by R. W. Alley

HarperCollinsPublishers

LEVEL
2

To Tom and Elynn and the design department, who have
worked so hard to make MathStart look so good
—S.J.M.

The illustrations in this book were painted with watercolor and
pen and ink on Strathmore bristol board.

HarperCollins®, 📖®, and MathStart® are trademarks of HarperCollins Publishers.
For more information about the MathStart series, write to HarperCollins Children's Books,
10 East 53rd Street, New York, NY 10022, or visit our web site at http://www.harperchildrens.com.

Bugs incorporated in the MathStart series design were painted by Jon Buller.

ANIMALS ON BOARD

Library of Congress Cataloging-in-Publication Data
Murphy, Stuart J., date
    Animals on board / by Stuart J. Murphy ; illustrated by R. W. Alley.
        p.   cm. — (MathStart)
    "Level 2."
    Summary: While driving her truck which carries a very special load, Jill counts and then adds the
number of animals that she sees on other trucks as they pass her on the road.
        ISBN 0-06-027442-5. — ISBN 0-06-027443-3 (lib. bdg.) — ISBN 0-06-446716-3 (pbk.)
        [1. Addition—Fiction.   2. Counting.   3. Animals—Fiction.   4. Stories in rhyme.]
I. Alley, R. W. (Robert W.), ill.   II. Title.   III. Series.
PZ8.3.M935An   1998
[E]—dc21   97-11733
                                                                 CIP
                                                                 AC

Typography by Elynn Cohen
            3   4   5   6   7   8   9   10

ANIMALS
ON
BOARD

My truck chugs slowly down the road
carrying a special load.

I'm the driver—my name is Jill.
I make sure that my load won't spill.

A big green truck goes roaring past
with three fierce tigers—count them fast.

7

Don't look now—here come two more.
Add them up and they might roar.

10

Six white swans are passing by.
They have wings, but they don't fly.

Look out now—here comes another.
Maybe it's their little brother.

13

What's this funny thing I see?
Four green frogs are passing me!

14

My truck keeps moving; it doesn't stop.
Four more frogs pass—hip-hop, hip-hop.

One more truck is passing now.
Seven prancing horses! Wow!

Then another truck I see.
This one holds another three.

The next truck holds a splendid sight—
nine plump pandas, black and white.

A red truck passes me with care.
What's in the back? Nothing's there!

But on the front's a special sign.
Time for me to get in line.

At last I've reached my final stop.
My truck held the tippy-top!

So now that all our work is done,
it's finally time to have some fun!

Can you find 5 tigers,
7 swans, 8 frogs,
10 horses, and
9 pandas?

32

FOR ADULTS AND KIDS

If you would like to have more fun with the math concepts presented in *Animals on Board*, here are a few suggestions:

- Read the story with the child and talk about what is going on in each picture.

- Ask the child to point to each animal as you count together. Ask questions throughout the story, such as: "If there are six swans and one more is added, how many swans are there in all?"

- Encourage the child to tell the story using his or her own words.

- Cut out animals from old magazines or catalogues and add them together. If you have two turtles and three dogs, ask, "How many animals are there in all?"

- Gather some stuffed animals. Arrange them into small groups that can be added together. Encourage the child to make up number stories about the animals. For example, "There are three bears and two bunnies. Five animals are going for a walk."

- Look at things outside the home—toys in a sandbox, items in a shopping cart, doughnuts on a bakery shelf—and practice adding them together.